WHEN THE OCEAN FORGETS

HOW CLIMATE CHANGE IS UNRAVELLING THE EARTH'S MEMORY

SHAMEEM KAZMI

Table of Contents

INTRODUCTION

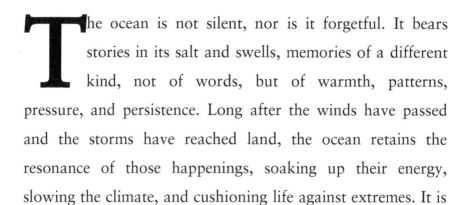

T he ocean is not silent, nor is it forgetful. It bears stories in its salt and swells, memories of a different kind, not of words, but of warmth, patterns, pressure, and persistence. Long after the winds have passed and the storms have reached land, the ocean retains the resonance of those happenings, soaking up their energy, slowing the climate, and cushioning life against extremes. It is an invisible, long-lived ability, which scientists have called "ocean memory". It is one of the most significant and least appreciated aspects of the Earth's climate system.

This book, *When the Ocean Forgets*, is an entreaty to step into that expansive, layered memory and learn its function and vulnerability. In science, narrative, and sustainability, this book considers how ocean memory has kept the Earth's climate stable for thousands of years and how that stability is weakening in the wake of climate change. While climate news is typically filled with stories of wildfires, retreating glaciers, and atmospheric carbon, the gentler, subtler narrative unfolding beneath the waves needs our immediate attention. This is the narrative of alteration not just in the ocean's health but in its capacity for remembrance and what that means for all of us.

The ocean has been the Earth's climate regulator for centuries. It absorbs more than 90% of excess heat due to greenhouse gas emissions, sequesters excess carbon, and transfers warmth and nutrients around the world in its intricate network of currents. It is not, however, an inert process. The ocean has a memory. It retains past conditions of sea surface temperatures, salinity, and heat content far beneath the surface, influencing the climate and the weather we see today and will experience in the future, like how a brain uses memory to inform decision-making, the memory of the ocean provides feedback into atmospheric patterns, impacting everything from South Asian monsoon cycles to hurricane formation in the Atlantic Ocean.

As with memory in the human brain, this memory of the world's ocean is subject to deterioration. When global temperatures increase, the ocean alters fundamentally in conditions on the surface and in its ability to store information. Current research, like that in *Nature Climate Change*, indicates that persistence in sea surface temperature

(SST) is decreasing. In plain language, the ocean is becoming less predictable. Where conditions previously conformed to familiar patterns of cool seasons, warm currents, and consistent circulation, there is increasing noise, disorder, and unpredictability.

When the ocean forgets, the consequences are profound. We are already witnessing the collapse of habitual seasonal patterns with calamitous consequences. Farmers can no longer rely on regular rains or planting opportunities; fisheries are witnessing unprecedented perturbances; coastal societies are subject to unprecedented storm surges and marine heatwaves that were previously unusual or unimaginable. Meanwhile, the deep sea is quietly sequestering heat in exchange. However, invisibly and not without effect, there is a gradual, mostly permanent, erasure of ocean memory that will define our climate for decades to come.

In this book, we will travel through the memory system of the ocean, from the surface down into the abyss. First, we learn what memory in the ocean is and how it functions, from thermal inertia through salinity patterns and the sluggish conveyor belts of ocean circulation. Then, we travel a little deeper and learn how this memory has facilitated the stability of the Earth's life by mediating climate fluctuations and maintaining biodiversity. Then, we see the weaknesses in the system: scientific findings that the memory of the ocean is breaking down. Along the way, we examine instances in which this shift already affects humans and ecological systems.

It is not, however, only a story of loss. It is a story of learning, both for humans and for others. In the face of fading memory, scientists are attempting to document, reassemble, and even recreate knowledge with technology. From the quiet floating Argo buoys to vast constellations of earth-observing satellites and machine learning algorithms, a worldwide campaign is underway to map and comprehend the altering mind of the ocean. It is not merely a series of technological achievements. It is a profound philosophical acknowledgement of the necessity of hearing the planet we are a part of.

Underpinning this account is a plea for sustainability in the sense of not merely cutting emissions or defending ecosystems but developing a relationship with the seaborne world on principles of humility, respect, and long-term thinking. While the ocean forgets, we are faced with the boundaries of our foresight. Our climate models, risk analysis, and economics are premised upon the expectation of a predictable planet. That presumption is starting to collapse. We need to adapt technologically, culturally, and ethnically.

To achieve this, we require not only scientists and policymakers but also citizens, educators, and communities to have ocean literacy. We need to understand the ocean as a living, remembering system to craft a future in which we survive and thrive. This requires reimagining education in a way that involves the ocean as a leading character in the climate narrative. It consists of creating policies and infrastructure equipped for a new kind of uncertain future. It consists of understanding that those least culpable for climate change, most notably communities in the Global South, are usually first and hardest affected by its impacts.

When the Ocean Forgets is a research-based but poetically charged investigation of these questions. It is written for the inquiring reader who does not merely want information but understanding, not data but meaning. Along the way, we will integrate research with stories, interweaving the sources from climate science, oceanography, Indigenous worldviews, and environmental thought. In doing so, this work aligns with the global vision of the United Nations

Sustainable Development Goals, particularly SDG 14 (Life Below Water) and SDG 13 (Climate Action), by contributing to a deeper awareness of oceanic systems and the urgent need to safeguard them.

This is the start of an immersion in a world that is, in its own way, still a mystery. We have mapped the seafloor with a far poorer resolution than that of the surface of Mars. We know less about the deep thermohaline circulation than we do about jet streams or atmospheric carbon. And still, the sea might be the answer to our shared climate future.

We cannot afford to ignore what is beneath. If the rhythms of the ocean need to shift, so too does our thinking on science, sustainability, and our place in the Earth's system. To save what memory is still there and increase resilience in the face of what is lost, we need to hear from the ocean, not just as a resource but as a teacher. This is an offering toward that listening.

Let's get started.

CHAPTER 1

THE OCEAN REMEMBERS: A HIDDEN INTELLIGENCE

The ocean does remember, but not the way humans do, with flashbacks of birthdays or forgotten tunes, but with something older, something deeper, namely, temperature, salt, velocity, and depth. Within the colossal network of water that blankets more than 70% of the planet, the past exists in a form that informs the future. This ability that scientists refer to as "ocean memory" is not a poet's exaggeration. It is a scientifically rooted notion that speaks of the capacity of the ocean to store knowledge of its condition and impact climate behaviour long after a perturbation has dissipated.

It requires starting with something basic about water: thermal inertia. Water resists sudden changes in heat, unlike land, air, or anything else. It requires considerably more energy to heat or cool the ocean than the atmosphere. This resistance builds persistence into a memory in that the ocean keeps an imprint of past temperatures and releases that heat gradually. This creates a lag in the climate system, an echo that resonates long after the actual sound.

Consider thermal inertia as the ocean's long memory. It won't forget in a hurry. When a heat spell increases the temperature of the ocean's surface, that heat does not just disappear with the setting of the sun or upon the passage of the next cold front. Instead, heat seeps into the top of the ocean, diffusing laterally and vertically, impacting weather systems weeks or months afterwards. Yesterday's heat can determine tomorrow's storm.

However, temperature is not memory. Salinity, the saltiness of the sea, is another essential variable. Salinity determines the density of the ocean water, and density controls its vertical and horizontal motion. Areas of greater salinity have the potential to become denser and sink, with fresher water rising to the top. This density-driven transport is the component of the ocean's vast conveyor belt: the thermohaline circulation. Like how neural circuits link distant regions of the brain, these currents interconnect far-flung areas of the world's oceans, carrying heat, nutrients, and momentum.

To some extent, the ocean is like a planetary brain. It receives input from numerous sources, such as solar radiation, river runoff, ice melt, and winds in the atmosphere and disperses that input through time and space in feedback loops. They are not dissimilar from cognitive systems' feedback processes: when change is detected, the system responds, compensates, or reinforces the signal, with a time lag typically involved. Atmospheric responses halfway around the world, for example, can be caused by the Pacific Ocean's surface temperature changes affecting Indian monsoons or Sahelian droughts. The memory of the ocean is global, recursive, and of long duration.

Just like a brain, the ocean is layered. The uppermost ocean, in contact with the atmosphere and weather systems, is the most dynamic and accessible for us to see. Below this is the thermocline, an active region with a sharp temperature gradient, and then there is the deep ocean, with its cold temperatures, sluggish movements, and longest memory. The processes that act upon the deep ocean, like long-term warming or the perturbation of the thermohaline circulation, can be imprinted for decades or even centuries. They are the ocean's slow memory, out of ordinary sight but essential in determining long-term climate conditions.

Take, for instance, the case of the El Niño–Southern Oscillation (ENSO). It is one of the most striking instances of memory in the oceans. In an El Niño occurrence, a build-up of warm water in the central and eastern Pacific upsets atmospheric circulation, impacting global weather patterns. The impact is felt as far apart as Peru, Indonesia, California, and East Africa. However, El Niño does not develop on its own. It builds up gradually, with indications in sea surface temperatures and beneath-the-surface heat content, months ahead of time. Scientists employ those indications in the memory of earlier states to predict the onset and strength of the next El Niño occurrence. Without memory in the oceans, those predictions would be impossible.

This intelligence is not faultless though. Like any system of memory, it can be disrupted. With the planet accelerating its rate of heat, the processes that sustain ocean memory are strained. Stratification, the division of ocean strata due to temperature and salinity differences, is growing stronger. Colder, denser water lies beneath warmer, less dense water, as vinegar lies beneath oil, cutting down on vertical mixing. This diminishing of mixing processes translates to the fact that surface anomalies in heat, carbon, or nutrients are less capable of reaching the deeper ocean, and, therefore, the ocean's capacity for absorbing and remembering such anomalies is weakened.

Meanwhile, circulation patterns are decelerating. The Atlantic Meridional Overturning Circulation (AMOC), a process that was compared by scientists like Charles Greene of Colby College in Maine to a conveyor belt that moves warm water from the equator up north in the North Atlantic and brings cold water back down in a southerly direction, has weakened. It is a pillar of the memory infrastructure of the ocean. If it decelerates dramatically as models indicate may occur this century, it will not just have implications for regional climate changes but for a fundamental diminishment in the ocean's ability to regulate itself.

We can envision this in terms of humans: think of a brain gradually losing its ability to create long-term memory, process stimuli logically, and respond with consistency. What you have is confusion, unpredictability, and, finally, dysfunction, and so it is with the ocean. When its memory fails, the climate system itself becomes more unstable. Rainfall patterns go wilder, heatwaves appear with increased suddenness, and seasonal cycles unravel.

However, there is something remarkable in the ocean's stability on a geologic timescale. The ocean has kept life buffered in a narrow range of temperature, chemistry, and motion through major extinctions, continental drift, and asteroid strikes. It has learned how to remain stable. It has learned, in a way, from billions of years of perturbation. This is why the time is so sobering: we are stressing a system that has withstood far greater forces, and we are stressing it faster than it can respond.

Science is beginning to comprehend the intricacy of this intelligence. Technology such as autonomous floats, underwater drones, and satellite altimetry is helping us monitor the ocean's memory in real-time. Readings from Argo floats, floating in the top two thousand metres of the ocean, indicate how heat and salinity are changing worldwide. Machine learning models now try to synthesise those measurements into predictions, feeding models that may be able to forecast changes in ocean memory and, by implication, our future climate.

Even our best models have difficulty representing the complete depth and complexity of ocean memory. This is because memory, in the ocean, as in the brain, is not fixed. It changes. It is dynamic, influenced by immediate circumstances and extended context. Ocean memory is not just stored data but instead is data processed, acted on, and recycled on scales of time and space. It is this dynamism that is its strength and its weakness.

Moving forward, we need to understand and respect this hidden intelligence for there to be sustainability. The ocean is not merely a sink for our emissions or a background for our economies. It is a remembering and living system whose health dictates the planet's destiny. If we disrupt its memory systems, we disrupt our own future and the systems we depend on.

In the following chapters, we'll observe how ocean memory engages with the systems we have come to rely on, such as agriculture, fisheries, and disaster preparedness, and how its destruction intensifies inequality, fragility, and doubt. First, we must listen intently to the shifts already underway beneath the waves. The ocean retains memory, but for how long is an open-ended question.

CHAPTER 2

THE MEMORY KEEPERS: HOW THE OCEAN STABILISES LIFE ON EARTH

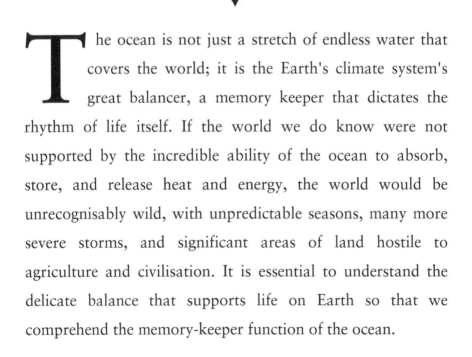

The ocean is not just a stretch of endless water that covers the world; it is the Earth's climate system's great balancer, a memory keeper that dictates the rhythm of life itself. If the world we do know were not supported by the incredible ability of the ocean to absorb, store, and release heat and energy, the world would be unrecognisably wild, with unpredictable seasons, many more severe storms, and significant areas of land hostile to agriculture and civilisation. It is essential to understand the delicate balance that supports life on Earth so that we comprehend the memory-keeper function of the ocean.

The process starts when the sun's rays reach the ocean's surface. The ocean, a vast heat reservoir, absorbs approximately 90% of excess heat created by greenhouse gas emissions since the industrial age. It's a remarkable task. Water's set of inherent physical characteristics, such as its high heat capacity, density, and circulation, means it can retain extremely high amounts of heat for extended periods, something that air or land, with their high rates of warming and quick cooling, cannot do. This ability of the ocean moderates global temperatures, reducing the severity of both the heat and the cold and ensuring a relatively consistent climate that supports life.

The memory of the ocean is inherent in this buffer function. The heat taken in is not merely passively stored in the top layer but seeps into deeper strata, where it remains for decades or years. This lag in the release of stored heat governs the rhythm and amplitude of seasonal variation and climatic patterns. For example, the relatively warm winters along the western seas of Europe and North America are mainly due to the heat carried by ocean currents from the tropics. The buffering of temperature extremes by the memory of the ocean avoids sudden climate changes that might otherwise decimate ecosystems and societies.

One of the best examples of the ocean's stabilising effect is its regulation of monsoon systems, upon which agriculture and freshwater availability for billions of humans rely. The South Asian monsoon, for instance, relies prominently on the temperature difference between the Indian Ocean and the Asian continent. The ocean keeps heat during summer that is gradually released to support the monsoon circulation. Ocean temperature and salinity patterns, aspects of ocean memory, directly affect the intensity and timing of monsoons. A weakened or delayed monsoon can result in drought and crop failure, whereas an exceedingly intense monsoon can result in devastating floods. The ocean's memory, in this manner, is a protector of seasonal predictability, ensuring food security for large numbers of individuals.

Aside from monsoons, the memory of the ocean also forms large-scale currents that support climate stability on Earth. The thermohaline circulation, otherwise known as the global conveyor belt, brings warm, equatorially sourced water near the surface, whereas colder, denser water descends and moves back in the direction of the tropics at increased depth. The circulation of heat around the globe keeps the planet's climate stable by ensuring that the tropical regions do not get unbearably hot while the polar regions are not entirely frozen.

The stability relies on the ocean's capacity to preserve the gradients in both salinity and temperature, two of its memory components.

One of the remarkable implications of this system is the relatively moderate climate of northwestern Europe compared to others of the same latitude. Without the balmy waters of the Gulf Stream and North Atlantic Drift, European winters would be considerably more severe, more akin to those experienced in regions of Canada or Siberia. Through salinity and thermal patterns, ocean memory directly maintains human civilisation by steadying regional climatic conditions.

The memory of the ocean is also a shock absorber for extreme events. It retains excess heat, limiting the rate and extent of atmospheric warming and, in turn, the frequency and intensity of heat waves and land-based droughts. Ocean currents also regulate the formation of storms, including tropical cyclones. The heat of the water near the ocean's surface controls the amount of energy available for intensifying storms; cold water lowers storm intensity, serving as a natural brake on storm intensification. Storm intensities would grow unmitigated without this shock-absorbing function, causing far greater damage to coastal settlements.

Furthermore, the ability of the ocean to sequester and recycle CO_2 links its memory with that controlling greenhouse gases. CO_2 is taken in by phytoplankton and other aquatic life during photosynthesis, and the deep ocean is a long-term reservoir for this carbon through the biological pump. Through sequestering carbon for up to several centuries, the ocean slows the rate of global warming, emphasising its function as a climatic and atmospheric chemistry stabilising force.

The memory of the ocean is not limited to chemical and physical processes but also shapes human history and civilisation. Ocean-driven climate patterns have determined the stability of agricultural societies, the rise and fall of civilisations, and even migration patterns. For instance, the predictability of the Nile's seasonal floods, upon which Egyptian civilisation thrived for thousands of years, was determined by Indian Ocean temperature-driven monsoon rains. Ocean memory interruptions and accompanying climate variability have, in most instances, preceded civilisation collapse.

In recent times, the predictability of ocean memory has supported contemporary agricultural planning and disaster preparedness. Seasonal forecasts primarily depend on the persistence of ocean conditions, such as the strength of currents and sea surface temperatures. They inform planting, water management, and disaster response decisions, saving lives and incomes globally.

Nonetheless, climate change is increasingly threatening the stabilising force of the memory of the ocean. Upon warming, stratification increases, decimating heat and nutrient mixing in the waters from the top and the deeper parts of the ocean. This compromises the heat and carbon absorption capacity of the ocean and its capacity for buffering. In addition, modifications in salinification patterns due to ice melting and precipitation changes alter the heat redistribution-relevant density-driven circulation.

Slowing of the Atlantic Meridional Overturning Circulation (AMOC), a prominent current associated with ocean memory, may provoke sudden regional climate changes with far-reaching implications.

The degradation of memory in the oceans destabilises the systems that societies rely upon, heightening the severity and incidence of intense weather events and undermining agricultural predictability. Rising sea levels and intensifying storms menace coastal societies, while monsoon dependent areas have increased food insecurity, affecting vulnerable people. The memory that protected civilisation is being eroded.

However, there is promise. New monitoring technologies, including autonomous Argo floats, satellite remote sensing, and underwater gliders, allow scientists to observe changes in ocean memory in greater detail. Combined with advances in climate modelling and machine intelligence, they promise greater forecasting of changes in climate patterns driven by oceans and the creation of adaptive approaches for sustainability.

Restoring the memory of the oceans will require joint efforts by all humans to mitigate greenhouse gas emissions, defend marine systems, and facilitate sustainable ocean management. Healthy oceans with undiminished circulation and mixing processes are necessary for maintaining the memory processes that regulate our climate and sustain life.

The memory of the ocean is a deep and valuable legacy. It has determined seasons, insulated extremes, and supported the forging of civilisation. To safeguard our future, we need to learn more of this hidden intelligence and act with purpose in its conservation. Only by valuing the ocean's function as the

Earth's great regulator can we create a sustainable world that is in harmony with its cycles.

CHAPTER 3

SIGNS OF FORGETTING: WHAT THE SCIENCE IS SHOWING US

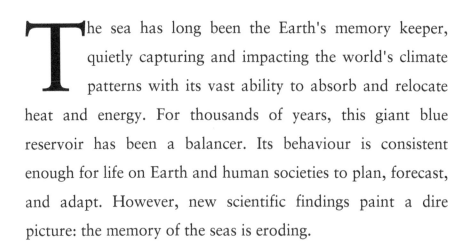

The sea has long been the Earth's memory keeper, quietly capturing and impacting the world's climate patterns with its vast ability to absorb and relocate heat and energy. For thousands of years, this giant blue reservoir has been a balancer. Its behaviour is consistent enough for life on Earth and human societies to plan, forecast, and adapt. However, new scientific findings paint a dire picture: the memory of the seas is eroding.

This is not a metaphor but a quantifiable shift in fundamental physical processes. It is an indicator of a reduction in the ocean's capacity for sustaining temporal temperature anomalies, maintaining steadiness in the circulation patterns, and therefore regulate the world's climate system. A flood of recent research, including groundbreaking research that appeared in *Nature Climate Change* in 2022, has revealed the signs of 'forgetfulness', unveiling a complicated and dynamic narrative with a serious consequence.

One of the most explicit signs of memory loss in the oceans is the fading of the persistence of surface sea temperature (SST). SST anomaly persistence is an essential gauge of how long the oceans can retain thermal information, affecting atmospheric patterns that influence weather, precipitation, and climatic predictability. In the past, when a part of the ocean warmed or cooled, those temperature changes persisted, propagating slowly and influencing climate conditions over weeks or months. This gradual, persistent propagation of heat created a temporal coherence, a memory trace, that imparted a type of inertia upon the climate system.

Despite this, satellite measurements, ocean buoys, and global climate models show significant reductions in this persistence in major ocean basins. The tropical Pacific Ocean, the setting for the El Niño–Southern Oscillation (ENSO), is a case in point. The surface of the Pacific Ocean warms or cools, driving changes in the atmosphere that impact rain patterns from Asia into the Americas, creating ENSO conditions. Over recent decades, there has been a reduction in the long-term persistence and amplitude of SST anomalies of ENSO, leading to increasingly erratic and unpredictable weather consequences. This has complicated forecasts, leaving farmers, governments, and disaster authorities less well-prepared for climate extremes.

Why is SST persistence declining? A principal reason is the growing stratification of the upper ocean layers due to global warming. With heat absorbed by the surface waters, the upper waters get warmer and less dense, creating a separate layer resistant to mixing with deeper, cooler waters. This stratification minimises the heat exchange of the ocean in the vertical direction, i.e., heat anomalies near the surface are less buffered and dissipate quickly. The ocean's thermal inertia, the epitome of its memory, is therefore undermined. Rather than a gradual heat diffusion, we experience faster oscillations, destabilising weather systems and undermining the coherence of seasonal patterns.

This diminishment of SST persistence is not limited to the Pacific Ocean. Analogous patterns are developing in the Atlantic and Indian oceans, interrupting local climate regimes that millions depend on. The Indian Ocean Dipole (IOD), impacting precipitation in East Africa and Australia, has manifested modified SST variability associated with ocean mixing and circulation changes. These interruptions in ocean thermal memory feed down to the land and impact crop failure, water scarcity, and enhanced exposure to both drought and flood.

In addition to SST persistence erosion, changes in ocean currents also serve as a grim warning of the ocean's memory in decline. Ocean currents serve as roads along which heat, nutrients, and salinity move, interweaving the global climate system. They are supported by the interplay of temperature and salinity, which help sustain their momentum and direction aspects of long-standing relationships of the ocean's memory.

One of the largest of these endangered currents is the Atlantic Meridional Overturning Circulation or AMOC. This enormous conveyor belt of warm tropical water heads for the North Atlantic, with cooler, denser water sinking and moving southward in the deep seas. AMOC is essential for regulating the climate of Europe and North America, allowing for the stable temperature that has enabled productive modern societies.

Recent findings that integrate satellite measurements, deep ocean sensor arrays, and high-end climate models indicate that the AMOC has slowed by around 15% since the mid-20th century. The slowdown is partly due to a mix of forcing by freshwater input from receding Greenland ice, increased precipitation, and warmer surface waters that upsets the fine balance between salinity and temperature that sustains the circulation.

The AMOC's weakening does not just alter ocean flows; it also heralds a failure of the ocean's memory to regulate climate. Slower circulation leads to reduced transport of heat north, resulting in potentially cooler temperatures in some areas of Europe and increased heat and dryness in the tropics. This variability can create feedback cycles, intensifying regional climatic extremes and leading to worldwide unpredictability.

The Indian Ocean and Pacific currents are also being disrupted. The Pacific Decadal Oscillation (PDO), a long-term variation of Pacific Ocean surface temperature impacting North American climate, has changed periodicity and amplitude, potentially associated with shifting ocean memory processes. Likewise, the thermohaline circulation of the Indian Ocean is less vigorous, impacting monsoon regimes essential for billions.

The overall result of these changing currents is a degradation of climate predictability. For decades, climate scientists and meteorologists have depended upon stable ocean patterns as the basis for decadal and seasonal forecasts. The memory of the oceans permitted forecasts of activities such as ENSO, monsoon intensity, and hurricane seasons. When those patterns begin to lose coherence, forecasting is less certain, and societies are left open to unpredictability.

Scientific research establishes that explained variance by ocean predictors in climate models has declined, especially in areas with the highest degradation of ocean memory. With this implication, the accuracy of climate projections, on which agricultural planning, water resource management, and disaster preparedness rely, is in jeopardy. The consequence is an increased disparity between societal needs and scientific ability for reliable predictions.

Furthermore, the degradation of ocean memory also heightens the likelihood of 'compound extremes', concurrent climate threats like heatwaves, drought, floods, and storms that overwhelm response systems. Lacking a stable memory, the timing and severity of such incidents become increasingly unpredictable, increasing their economic and human toll.

The declining ocean memory is also expressed in altering biogeochemical cycles. For instance, current-driven and vertical mixing nutrient upwelling supports fisheries and marine ecosystems worldwide. Impeding these cycles by disrupting circulation patterns has a domino effect on the biodiversity of the seas and the coastal community's economy.

It is necessary to note that although these changes are worrying, our insight is still developing. The ocean is an intricate, adapting system with feedback that may help reduce or accelerate memory loss. For example, excess heat in deeper water could sometimes ultimately return to the top, recovering a degree of persistence. In contrast, swift ice melt and freshwater addition could boost memory decay. Current research, especially with advanced climate models and improved observation arrays, seeks to disentangle these processes.

One of those efforts is the worldwide Argo programme, with thousands of autonomous floats that continuously measure temperature, salinity, and currents in all the world's oceans. Combined with satellite measurements, these data streams enable scientists to monitor changes in indicators of ocean memory in near real-time. Machine learning and artificial intelligence also facilitate the incorporation of large datasets to refine knowledge and forecasting of the evolving state of the ocean.

Nevertheless, the core challenge still exists: the ocean's ability to keep the Earth's climate in balance is being pushed harder than ever. Human activities, such as greenhouse gas emissions, deforestation, and pollution, are driving the ocean to tipping points where its memory mechanisms can be irreversibly changed. The impacts are worldwide, crossing borders and influencing all corners of life.

Identifying the signs of forgetting is a wake-up call. It emphasises the need for emissions reduction and the health of the oceans to protect those processes that sustain climate stability. It also points to the need for resilience planning in exposed communities, with planning that includes uncertainty and variability.

Long taken for granted, the memory of the ocean now appears as a delicate treasure that holds the secret of the planet's future climate. The science is indisputable: we are on the early watch of its deterioration. What we do next is up to the decisions we make today, whether to pay homage and defend this hidden wisdom or lose it and, in doing so, lose the stability upon which all of life relies.

CHAPTER 4

FROM RHYTHM TO CHAOS: THE END OF RELIABLE SEASONS

For much of human history, the cyclical rhythm of the seasons has been the planet's most constant metronome. The periodic veering between wet and dry spells, cold snaps and heatwaves, monsoon rains and harvest have governed everything from ancient agricultural calendars to the movement of animals and the layout of civilisations. We have grown up in harmony with these patterns, depending on them to plant crops, time rituals, prepare for storms, and allocate water. However, more and more often now, that rhythm is breaking. The seasons are unreliable, moody, and, in some regions, barely recognisable.

At the core of all this disruption is something both deep and underestimated: the loss of ocean memory. In earlier chapters, we discussed how the sea's ability to capture and share heat, its thermal memory, lies at the root of global climatic stability. When its memory falters, the cascading consequences fall from the realm of abstraction into failed rains, premature frosts, sea die-offs, and amplifying extremes. The sea, once the great conductor of the Earth's climatic rhythm, is now part of causing chaos.

The narrative starts with sea surface temperature, which is the temperature of the sea. When ocean memory is high, sea surface temperature anomalies last. A hot area of sea off Peru's coast, for example, may remain for months, offering the atmosphere a steady signal and enabling scientists to project downstream impacts like the arrival of El Niño or alterations in the Indian monsoon. As the ocean's memory decays, the anomalies disappear more rapidly or vary in unpredictable patterns. Without this steadiness, the atmosphere gets conflicting signals. The outcome is increasing unpredictability of rains, wind behaviour, and seasonal transitions.

This unpredictability is not merely of scientific interest; it is profoundly disruptive on the ground. Think of the Sahel of Africa, a semi-arid belt across the continent where life hangs on the timing and reliability of the rains. Traditionally, the arrival of the rains follows in a predictable fashion from the control exerted by sea surface temperatures in both the Atlantic and Indian oceans. When temperatures vary coherently and sustainably, the monsoon similarly varies in response.

But in recent years, people in the Sahel have experienced erratic rains, seasons that start too early, too late, or suddenly in bursts of destructive power instead of in steady, life-giving streams. These have produced poor crops, interrupted planting times, and food shortages for tens of millions of people. Research attributes the instability to changes in ocean circulation and SST persistence. When the ocean has an unreliable memory trace in which to reside, the atmosphere is an unstable partner with sudden and sometimes violent mood changes.

The Indian subcontinent is another sobering case in point. For countless generations, the Indian monsoon has come with note-perfect timing, an experience that has conditioned everything from rice farming to religious festivals. In the past twenty years, however, the monsoon has become ever more unpredictable. It is occasionally late, bursting in for dramatic effect in torrents of water and deluge and occasionally failing at critical stages of the growing season.

In 2019, entire regions of India went through drought and flood in the same year at times in the same area. Farmers in Maharashtra, for example, sowed seeds after what seemed like the rains had begun, only to see them dry out as the rains disappeared. Weeks later, deluges destroyed the topsoil and nascent crops, levelling what was left behind. These wild swings are associated with changing SSTs in the Indian Ocean, alterations in the Indian Ocean Dipole, and diminishing thermal gradients, all related to the loss of ocean memory.

The disruption is not limited to the tropics. Even Europe, buffered as it was in the past by the relatively consistent thermal behaviour of the Atlantic, has experienced record weather extremes. Germany and Belgium experienced floods in the summer of 2021 following a stationary weather system depositing too much precipitation upon saturated ground. On the other hand, Southern Europe was parched in a record heatwave during the same period as temperatures in Sicily reached close to 49°C. These extremes align with what scientists have dubbed "weather weirding" as atmospheric behaviour lessens in its attachment to long-established seasonality. A contributing factor is the slowing of the Atlantic Meridional Overturning Circulation (AMOC), described in previous chapters. When this vital conveyor belt of heat decelerates, temperature gradients between the tropics and poles decline, changing the jet stream and disrupting weather systems. Instead of the orderly eastward march of weather fronts, we experience them becoming stuck or wandering, resulting in extended droughts or rains in the same area.

In the meantime, the sea's thermal chaos is not just felt on the surface. Beneath the waves, ecosystems are being disrupted by rapid and sometimes sudden changes in temperature, known as marine heatwaves. These previously uncommon and temporary affairs are becoming more frequent, longer lasting, and intense. In 2016, an unusual marine heatwave known as "the Blob" formed in the northeast Pacific and caused shockwaves in the seas' food chains. Kelp forests off California's coast died back. Fish populations fled to colder waters and left fisheries in disarray. Seabirds starved as their food disappeared. These impacts were directly attributed to persistent yet anomalously high sea surface temperatures, the onset and duration of which broke all previous records.

Ocean memory loss changes not only the temperature but also the timing and distribution of important ocean currents. These changes impact the upwelling areas where cold, nutrient-rich waters ascend to the sea's surface, sustaining rich ecosystems and productive fisheries. On the coasts of Peru and Namibia, altered current patterns have undermined upwellings, causing ripple effects on fishery populations, seabird communities, and coastal economies. The life pulse of these ecosystems is breaking down just as it is on land.

What makes it so disquieting is the rapidity of the shift from rhythm to chaos. Even the most advanced climatic models did not forecast most of the changes in the last ten years. One of the reasons is that the models had been tuned using a climatic history where ocean memory was strong. When that memory weakens, the fundamental assumptions of climatic forecasting must be revised. We're moving into an era where what happened in the past is less certain to guide future conduct, a shift not just for science but for the ability of society to prepare and react.

Agricultural planning, water management, public health, and infrastructure design all rely to some extent on seasonality predictability. When predictability disintegrates, the efficiency of institutions deteriorates. Low-income countries and poor communities stand at grave risk as they are unable to hedge against climatic unpredictability. In most instances, such communities find themselves in the harsh paradox of contributing least to climatic changes, diminishing ocean memory yet suffering the most from its repercussions.

To add complexity to the situation, the breakdown of seasonality does not necessarily play out as it does at every point on the planet. In some places, climatic patterns can seem reassuringly steady until some threshold passes. The Amazon basin can experience 'average' rainfall for years before suddenly lurching into extended drought or deluge due to distant oceanic perturbations. These sudden changes can be preceded by slight alterations in SST gradients or circulation patterns, again emphasising the need for high-resolution observation of the oceans.

A promising aspect in this context is the availability of sophisticated ocean observing systems, including Argo floats, remote-sensing satellite radiometers, and ocean gliders to deliver near-continuous temperature, salinity, and current behaviour measurements. Coupled with AI-aided forecasting models, these instruments can identify precursors to the collapse of seasonality. Even the best instruments, however, cannot restore an erasing memory. We can, at best, steer through the maelstrom with enhanced consciousness.

In the end, the disappearance of ocean memory signals planetary change, a departure from the slow, understandable cycles of the world to one of sudden, discontinuous change. The loss of dependable seasons is not just an issue of meteorology; it is an affront to the foundation of human culture, which has been inextricably bound to the rhythm of nature. As we face this new reality, we are called upon to shift material circumstances and change our way of thinking. The comfort of predictability is being replaced with the new state of living in uncertainty. Therein, resilience ceases to be the preservation of the established rhythms at all costs but learning to move with the new rhythms, however jagged and uncontrolled they might be. The ocean does not act out of meanness in its forgetting. It responds to pressures we have put on it, interpreting more than 90% of the extra heat trapped by greenhouse gases, tolerating acidification, stratification, and changed salinity flows. It does what systems in trouble do: it rearranges. Whether we perceive this rearrangement as chaos or discover new patterns of coherence in it is, in part, a function of how fast we act. The clock continues to tick, but the beat has changed.

CHAPTER 5

THE DEEP FORGETTING: WARMING BELOW THE SURFACE

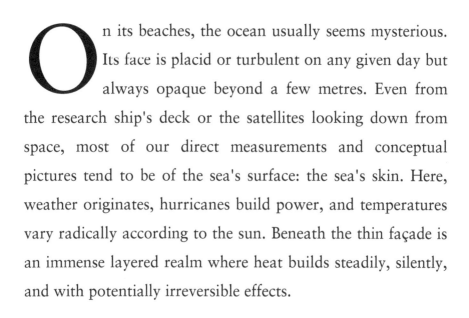

On its beaches, the ocean usually seems mysterious. Its face is placid or turbulent on any given day but always opaque beyond a few metres. Even from the research ship's deck or the satellites looking down from space, most of our direct measurements and conceptual pictures tend to be of the sea's surface: the sea's skin. Here, weather originates, hurricanes build power, and temperatures vary radically according to the sun. Beneath the thin façade is an immense layered realm where heat builds steadily, silently, and with potentially irreversible effects.

This chapter enters that submarine reality: the domain of deep ocean warming. If ocean memory is a metaphor for the climatic system's ability to "remember" previous conditions and modulate future ones, then the deep ocean is its long-term repository. However, that repository is being altered. Those layers of the sea, which previously stored the equilibrium of planetary heat and stability, are now consuming extra energy at record levels. This deep forgetting is long-lasting, silent, and largely invisible and could be one of the most significant changes of our era.

To grasp the seriousness of deep ocean warming, we must start at the fundamentals of where the ocean accumulates heat. About 93% of the extra heat from anthropogenic greenhouse gases has gone into the ocean. Of that, an ever-growing share is being held at depths of more than 700 metres, a region not visited much by seasonality or short-term weather. This redistribution has been a planetary mercy, keeping the atmosphere from more extreme warming and effectively buying time for us, but at what price?

In contrast to its surface waters, the deep ocean is slow to respond to change. Polar-produced water masses, such as the Antarctic Bottom Water or the North Atlantic Deep Water, sink and move along the ocean floor at their leisure for centuries, or even thousands of years to complete their circulation cycles. When such deep currents get loaded with heat energy, they are stored away, creeping through the global system at a time scale beyond human experience. We may not experience its full effects for generations, but it's impossible to undo once entrenched.

This thermal trap has profound implications for how we model climate dynamics. On the surface, it can seem reassuringly steady. A mildly warming year can be read as forward momentum or stagnation when the deep ocean stores energy in a sponge-like way. The phenomenon can cause what scientists refer to as "warming in the pipeline", an inevitability of eventual rise in surface temperature if emissions were to be curtailed right now. The heat is in the system; it has just not yet reached the surface.

Recent studies have indicated the amount of energy being built up at depth. Publications, including *Nature Climate Change* and *Geophysical Research Letters*, have reported that oceanic heat in the abyssal depths (over two thousand metres) has surged in the last two decades. In the Southern Ocean, particularly surrounding Antarctica, where deep waters form, subsurface temperatures are rising faster than predicted. The heat is being transported laterally and vertically, altering the thermal profile of the sea in manners that disrupt its classical memory function.

Why does it matter for climate stability? For one thing, it's the loss of vertical stratification. Under natural circumstances, the ocean is stratified, with denser, colder water at deeper levels and warmer water at the surface. This stratification works as a barrier permitting the concentration of heat at the sea's surface, where it can be released slowly at its own pace or exchange with the atmosphere. As the deep ocean becomes warmer, the structure is lost. Warmer water moves into the deep waters, diminishing the contrast in density and suppressing the mixing of the sea vertically.

The consequence is the double bind. On the one hand, stratification on the surface heightens, reducing the ability of the higher levels of the ocean to release heat into space, resulting in more severe marine heat waves and on-surface anomalies. On the other hand, where mixing takes place like during storms or upwellings, it transports warmer deepwater into the near-surface ocean, heightening atmospheric warming. In either case, the deep forgetting establishes feedback loops that undermine the ocean as a global climatic stabiliser.

Another effect is sea level rise. Most people think that melting ice is the cause of rising seas. Thermal expansion is just as powerful an impetus. When water is heated, it expands, occupying more space without changing mass. Warming of the deep ocean directly contributes to the expansion. Because it occurs deep down and progresses slowly over tens of years, it adds a long-term, irreversible factor in sea level rise. Even if all the ice sheets and glaciers were somehow stabilised, thermal expansion due to the warming at depth would drive sea levels upward for another few hundred years.

In the western Pacific and Indian Ocean regions, where deep waters have had time to accumulate vast quantities of heat energy, regional sea levels have altered well more than the global mean. Low-lying countries are especially at risk not only from advancing waters but also from the destabilisation of subsurface environments upon which other marine life and weather patterns rely.

The environmental consequences of deep ocean warming are no less serious. Several deep-sea animals are tightly calibrated to small thermal tolerances. A shift of just several tenths of a degree changes metabolic levels, reproduction patterns, and survival limits. Coral reefs, already under pressure from surface-level bleaching exposures, rely on the input of cold, nutrient-laden water from the deep. As it warms, its ability to cushion surface heating diminishes, resulting in longer lasting and more severe bleaching.

In addition, the biological pump, the mechanism through which the ocean locks away carbon as organic matter sinks, depends on steady thermal gradients. When the deep ocean becomes warmer, the pump's effectiveness can dwindle, decreasing the ocean's capacity to function as a carbon sink. This feeds back into the larger climate system again, decreasing the ocean's capacity to counteract emissions and increasing atmospheric warming.

We effectively erase the ocean's memory not merely through surface churning and changing currents but by fundamentally rearranging the thermohaline structure of the deep sea. Because the deep ocean functions on multi-century time scales, the consequences of forgetting are not easily reversed; they are an implicit long-term planetary bet, the full implications of which will materialise across lifetimes and epochs.

Still, for all the severity of the transformation it portends, deep ocean warming is underappreciated by the public and underestimated in policy. It is intangible and abstract and does not possess the visceral imagery of melting ice cliffs or raging wildfires. There has not been a viral photo of abyssal warming, no crashing ice cliffs, no blazing forests. It is a crisis of accumulation, not a spectacle.

Maybe that's why it requires us to be aware. It reminds us that some threats don't scream, and others whisper insistently from beneath.

Researchers are competing to chart and explain this silent turbulence. Deep Argo float deployment has started to offer salinity and temperature readings to six thousand metres. These underwater robots give us an unprecedented view of the unseen workings of the planet's climatic system. Together with ocean gliders, mooring instruments, and satellite altimetry, they are putting together a more detailed picture of the movement of heat through the ocean across time.

Still, the task is not an easy one. Monitoring the deep ocean is expensive, technically challenging, and logistically complicated. The sheer extent of the abyssal plains, the inaccessibility of crucial formation regions, and the sluggishness of change all plot against the task as being of monumental proportions. Unless we have these measurements, our models of the Earth's climate will continue to underestimate the planet's warming momentum, and our strategies for reducing it will remain in disarray.

It is also an ethical issue. The deep ocean has too long been viewed as a dumping site or an enigma, an out-of-sight space and hence out-of-care. As we come to appreciate its core function in controlling the Earth's climate and safeguarding its memory, we need to develop a new story of stewardship, not exploitation.

Conservation of the deep sea is not just about maintaining its biodiversity. It is also about recognising it as a thermal reservoir, carbon sink, conveyor of currents, and planetary archive. It recognises the health of our atmosphere, coastlines, food systems, and future as inextricably connected to what occurs two thousand metres beneath the waves. The deep forgetting is not sensational. It does not declare itself in thunder and fire. It happens as quietly as the sea rising across an interval of millennial time. Its power is in that silence. Behind it lies the inexorable, invisible piling of heat, the realignment of the ocean's interior, and the gradual loss of one of the Earth's most vital memories. We do it at our own risk. In forgetting the deep sea, we may be assisting it in forgetting us.

CHAPTER 6

GHOST CURRENTS: DISRUPTION IN THE ATLANTIC AND BEYOND

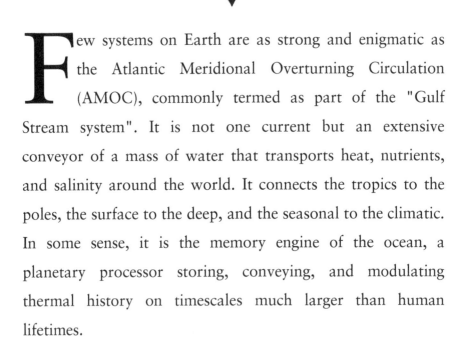

F ew systems on Earth are as strong and enigmatic as the Atlantic Meridional Overturning Circulation (AMOC), commonly termed as part of the "Gulf Stream system". It is not one current but an extensive conveyor of a mass of water that transports heat, nutrients, and salinity around the world. It connects the tropics to the poles, the surface to the deep, and the seasonal to the climatic. In some sense, it is the memory engine of the ocean, a planetary processor storing, conveying, and modulating thermal history on timescales much larger than human lifetimes.

Today, this essential system is in serious decline. Once a reliable rhythm, pulsating in sync with Earth's seasons and maintaining the equilibrium of climatic zones, it now stutters and falters. The AMOC is decelerating; its once strong pulse has grown weak and irregular. Traces of currents past can be seen but are vanishing, withdrawing into a ghostly instability capable of redrawing the contours of climatic predictability.

To know what's happening and why it's significant, we need to understand the AMOC. Simply put, the AMOC is an extensive ocean circulation system that carries warm salty water from the tropics northward at the surface of the Atlantic Ocean. When it reaches higher latitudes, it cools, gets denser, and plunges into the deeper sea before it travels southward at depth and ultimately upwells in the Southern Hemisphere or the Indian and Pacific Oceans. This global conveyor belt transports heat from the equator to the poles and is pivotal in maintaining regional climates, especially in North America, Europe, and West Africa.

This circulation is fuelled in part by variations in temperature and salinity thermohaline forcing, as oceanographers refer to it. Warm water is less dense than cold water, and saltier water is more dense than fresh water. When these attributes change in significant enough proportions, they can break the density gradients that fuel the AMOC's power. Freshwater inputs from melting ice sheets, especially Greenland's, in the last century, plus rising surface temperatures and precipitation, have diluted the salinity and changed the thermal structure of North Atlantic waters. These have weakened the downward sinking movement, which gives the AMOC its power.

Recent research in publications such as *Nature Climate Change* and Nature Geoscience indicates the AMOC is nearer to its critical tipping point than we have assumed. By extracting proxy records from sediment cores, coral growth bands, and ice cores, scientists have stepped back through time to reconstruct the history of the AMOC for the last thousand years. What we discovered was alarming: the present-day weakened state is probably the largest in more than one thousand years. Satellite records and ocean buoys agree that it's not just variability; it's something structural in the ocean's circulatory system.

The consequences of an AMOC slowdown are enormous and potentially calamitous. In Europe, a weakened AMOC would bring colder winters and warmer, drier summers, paradoxical but understandable via altered atmospheric circulation. In West Africa, it would transform the African monsoon system, diminishing rainfall and endangering agricultural subsistence for tens of millions. Reduced moisture transport in the Amazon would push the rainforest towards dry-state collapse. On the North American east coast, sea levels would rise much more quickly because of localised changes in the ocean currents and gravitational rebound on the continents.

Still more alarming are the knock-on effects, a ripple of consequences triggered by the destruction of this oceanic memory. The AMOC does not operate in isolation. It is inextricably interconnected with the Intertropical Convergence Zone, the El Niño–Southern Oscillation, the Indian monsoon, and polar ice behaviour. Upset in one sector of the system can have chaotic repercussions elsewhere. Ocean memory, under pressure from surface warming and deep ocean heat absorption in the first place, is broken and distorted, much as the synapses of the mind become disordered.

This breakdown can be conceived not as merely a physical process but as a failure of predictability. For generations, the relative reliability of the AMOC has enabled civilisations to base expectations on their farming, economic, and social systems. Crops had been sown in reliance on rainy season rains; fishing routes had tracked established routines of nutrient upwelling; cities had been established where the climate appeared to be timelessly temperate. As the AMOC falters, all these assumptions fall. Past norms become unreliable. The future has lost its anchors.

This is not just theoretical. We can see evidence of today's impacts. The purported "cold blob" in the North Atlantic, a patch of anomalously cold sea surface temperature south of Greenland, has been attributed to the AMOC slowdown. Although global sea surface temperatures have increased, this region continues to be anomalously cold, indicating less heat being transported by the AMOC. Paradoxically, the surrounding area has seen record heatwaves and odd weather patterns, testifying to the planet-stabilising ability of the AMOC.

Sea ecosystems are also being disrupted. In the North Atlantic, species adapted to steady circulation patterns are changing their range. Cod, herring, and mackerel are migrating in pursuit of cooler waters. These changes have implications for fisheries management, food security, and international maritime treaties. Jellyfish and invasive species thrive in some of these altered environments, in some cases turning formerly productive fishing grounds into environmental wastelands.

As with deep ocean warming, the full effects of disruption in AMOC are also likely to be slow in coming. The system's inertia means it can take decades of continued decline even if carbon emissions are drastically reduced today. This phenomenon in climatic modelling is referred to as "hysteresis", the notion that systems don't necessarily go back to their original state if forcing is removed. Once a threshold has been reached, the way back can be slow, uncertain, or impossible.

The worst-case fear is the entire collapse of the AMOC. This would not imply the disappearance of ocean flows but instead a wholesale realignment of global circulation. We have evidence from history to suggest it has occurred before. During the Younger Dryas era some twelve thousand years ago, an enormous input of freshwater into the North Atlantic was suspected of temporarily shutting down the AMOC and plunging regions of the Northern Hemisphere into a millennium-duration cold snap. A complete repeat of anything like this remains uncertain, but models suggest that a collapse could be triggered if global warming exceeds certain threshold levels, possibly in the range of 2 to 4°C above pre-industrial levels.

Such a collapse would have profound implications for ocean memory. If the AMOC did not exist to transport heat and salt through the Atlantic basin, stratification would become more pronounced, amplifying the feedback, and regional climates would diverge radically. The ocean's buffering and balancing capacity for the climate system would be greatly undermined. Essentially, one of the world's key memory banks would be erased.

What can then be done?

AWARENESS. For all its significance, the AMOC is an unfamiliar idea for many beyond scientific and policy communities. Making people understand its applicability and how an unseen current can impact Ghanaian farming, New York City's flooding, or the Amazon's droughts is crucial to generating the political will to act on climate change. Ocean literacy must be more than an addendum to environmental studies; it must be at the core.

Second, monitoring. Recent initiatives like the RAPID array, a string of moored sensors across the Atlantic from Florida to the Canaries, have given us precious records of the behaviour of AMOC since 2004. More is required. We must extend ocean observing networks, fund autonomous sensing systems, and fund long-term research. The ghost currents will not yield their secrets willingly. Only continued, high-resolution observation will let us identify warning signs of a tipping point before it is too late.

Third, emissions reduction. Although ocean systems are vast and sluggish, they are not beyond our control. For every tonne of carbon dioxide emitted into the global atmosphere, it adds to the thermal load on the ocean, enhances ice melt, and alters salinity gradients. Restricting warming to 1.5°C according to the Paris Agreement can be our best hope of maintaining the integrity of the AMOC and its memory functions.

Ultimately, adaptive resilience. Even as we try to stabilise the climate, we need to prepare for the prospect of some of the changes being irreversible. That will require us to develop resilient farming systems, redesign sea-level infrastructure, rethink fisheries management, and prepare for migration caused by the climatic shift. If the ghost currents keep dying out, we must learn to coexist in a world where the contours of stability have changed.

In a very real way, the history of the AMOC is one of planetary memory and potential loss. These massive currents are not just physical currents of water; they are channels of history, storing previous centuries' thermal and salinity signatures and sculpting the coming years through their slow, steady movement. Weakening them is not just something of scientific interest. It's an earth-shaking change. To lose the AMOC is to lose one of the Earth's most ancient rhythms. It is to relinquish one of the circulatory intelligences that have supported our climates, cultures, and faith in the season's reliability. It is to open ourselves to the unknown, not through explosions of dramatic impact or cataclysms of sudden destruction but through the slow hollowing out of an ancient and deep voice. The ghost currents still run, whispering across the Atlantic as echoes of a process remembering more than it ever forgot. Whether we can restore them or if we're well into the wake of their decay is one of the fundamental questions of the day.

CHAPTER 7

LIVING IN THE UNKNOWABLE: RISKS TO PREDICTION AND PLANNING

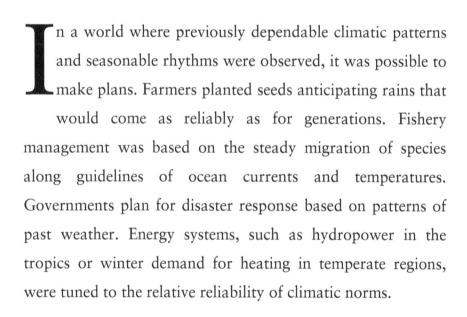

In a world where previously dependable climatic patterns and seasonable rhythms were observed, it was possible to make plans. Farmers planted seeds anticipating rains that would come as reliably as for generations. Fishery management was based on the steady migration of species along guidelines of ocean currents and temperatures. Governments plan for disaster response based on patterns of past weather. Energy systems, such as hydropower in the tropics or winter demand for heating in temperate regions, were tuned to the relative reliability of climatic norms.

Today, that basis is eroding. The idea of ocean memory, the recurrence of sea surface temperature, salinity regime, and circulation couplings breaks down. The outcome is not just a warmer planet but one less fundamentally in control. We are moving towards a more variable climate where statistical probabilities become irrelevant, and planning for harvests, fish catches, infrastructure development or emergency management is increasingly at risk.

The deterioration of ocean memory is at the heart of this unpredictability. When the ocean's surface temperature decreases from season to season, ripple effects work their way through atmospheric systems. Oceanic variation forces much of the season and interannual weather patterns, including monsoons, droughts, storm tracks, and heat waves. When the ocean forgets, the atmosphere stumbles in its dance and the resulting weather is disjointed, erratic, and, more and more often, treacherous.

Already, the effects are being keenly felt in agriculture. Crops require reliable rainfall and temperature patterns, and many have fine-tuned themselves to very specific season signals. Yet, in the last few years, farmers in sub-Saharan Africa, South Asia, and Central America have found themselves out of rhythm with the climate. Rains arrive too early, too late, or not at all. Dry spells last for weeks, and then sudden torrents of rain wash away the precious topsoil. Heatwaves occur during pollination stages and crush yields.

These are not one-off exceptions; they are symptomatic of growing unpredictability at the systems level. Disconnection of climatic patterns from past norms has serious implications for food security. Oral and written calendars become redundant. Climatological mean-based agricultural models are stretched too thin. Crop insurance plans struggle to price risk as governments rush to react to regional failures.

Fisheries also suffer from the turbulence. Not only are ocean temperatures warming, but they are also destabilising. Essential spawning areas, nutrient-rich upwellings, and migratory channels respond unpredictably. Fish react to minor thermal shifts by changing their habits and moving hundreds of kilometres in search of appropriate conditions. Entire ecosystems realign and break established trophic relationships and invalidate management plans predicated on archaic presuppositions.

For many coastal communities dependent on small-scale fisheries, such unpredictability means livelihood instability. Boats come back empty. Seasonal anticipation fails. New fishing grounds lead to new conflicts. With satellite tracking and extended range, industrial fleets can better pursue the migrating resource frontiers, increasing inequality in the sector. Without credible forecasts, fisheries management is reactive and too slow to curtail overfishing or ecological meltdown.

The unpredictability of the climate also attacks disaster response systems. Tropical cyclones, previously seasonal in timing and geographically bounded in their occurrence, are now organising beyond conventional "windows" and impacting unprecedented regions. The 2020 Atlantic hurricane season was record-shattering in terms of the number of storms and their erratic behaviour and rapid intensification features becoming ever more correlated with anomalous oceanic temperatures and disturbed memory patterns.

Once storms no longer obey earlier patterns of occurrence or magnitude, readiness is undermined. Evacuation routes, shelter space, and floodplain mapping are all predicated on the assumption that the future will resemble the past. But ocean memory decays now, and the assumption does not apply. In such circumstances, disaster risk reduction is more of an improvisation exercise than a product of planning. Warning systems must be modified to accommodate narrower windows and less precise forecasts. Infrastructure must be built not for known hazards but for unknown hazards.

The energy sector also stands at risk. Low-carbon systems like hydro, wind, and solar are weather-dependent. Unpredictable rainfall impacts reservoir levels; changing wind directions impact productivity from turbines; extended cover from clouds interferes with solar production. In a steady-state climate, these changes would be predictable and offset. In an unstable one, energy planning is guesswork. Grid reliability is lost. Load forecasting ceases to be an accurate tool. Confidence in investment evaporates.

Furthermore, energy demand itself changes due to climatic variability. Unseasonable cold snaps increase the demand for heating, and extended heatwaves boost air conditioning usage. These sudden and dramatic cycles put pressure on infrastructure not designed to handle such extremes. In developing countries with already tenuous grid systems, this imposes another vulnerability.

Underlying all these sectoral impacts is something more profound: we're pushing our climate models to the breaking point. Those models, the foundation of climate science and policymaking, rely on statistical continuity and boundary conditions. These models are incredibly complex and advanced but are also constructed to model systems with some stability. When stability is lost, ocean memory wears off, and amplifying feedback loops start to occur, the models become doubtful sources of reliable prediction.

This is not a modelling science failure but an expression of the challenge. Modelling a chaotically behaving climatic system where driving forces vary in magnitude and behaviour is qualitatively different from simulating a warming but steady world. Uncertainty bands broaden. Ensemble forecasts diverge, the range of conceivable futures increases, as does the complexity of policy and planning.

Ironic as it is, the more uncertain the climate, the more we need accurate predictions. Policymakers, city planners, farmers, and investors rely on forecasts to inform long-term choices. The insurance world is especially confronting the irony. Risk is no longer something that is simply priced based on loss experience. Disastrous occurrences are cropping up beyond the limits of statistical probability, frequently with calamitous financial impacts.

In response to these heightened risk exposures, some insurers are abandoning high-risk zones altogether, coastal regions with a high risk of flooding or storm surge. Others are charging premiums to unaffordable prices or limiting coverage. This loss of insurability has consequences well beyond their policyholders. It imperils the social contract of insurance: shared protection against the unknown. As the social contract unravels, inequality widens. Those who can buy customised protection or exit from risk-prone areas do; those who cannot are left vulnerable.

The expression "living in the unknowable" is not an exaggeration but an account of an emergent reality. In the Anthropocene era, human beings have become a driving force in shaping planetary systems. These systems are, in turn, changing what we can know and respond to. Interactions among climate change, oceanic memory loss, and socio-economic exposure generate what some academics call "deep uncertainty" circumstances under which even the risk itself is in flux.

How do we prepare for the unimaginable?

The starting point is to recognise that uncertainty is not just a temporary situation to be overcome but has become an ongoing characteristic to be monitored. That requires a new planning paradigm of flexibility, redundancy, and adaptive capacity rather than optimisation and efficiency. In agriculture, this may imply sowing crop varieties with resilience to weather patterns, investing in water storage, and diversified income. In fisheries, it requires adaptive co-management, shared real-time information, and cautionary quotas. In city planning, it means designing infrastructure for modularity and climatic buffers.

Second, we need to invest in improved observation. The ocean, and particularly the deep ocean, is still under-sampled. Although systems such as the Argo floats and remote sensing from satellites have transformed oceanography, significant gaps still exist, most notably in boundary currents, polar oceans, and subsurface thermoclines. We need real-time global-scale, high-resolution, open-access data to chart an uncertain climate.

Third, we must reshape the way we communicate about climate change. Uncertainty should not be conflated with ignorance and inaction. Instead, it can be framed as an arena for prudence, resilience, and creativity. We know people understand the unpredictability of life; what we need are the instruments, stories, and institutions to help them act wisely in this situation.

Taken together, these actions speak directly to the ambitions of SDG 13 (Climate Action) and SDG 14 (Life Below Water), which call for strengthened resilience, adaptive planning, and enhanced knowledge of marine ecosystems. Responding effectively to oceanic uncertainty is not just a scientific imperative, it is a global development priority.

Ultimately, we must adopt systems thinking. In an interconnected world, the separation of knowledge into agricultural science, meteorology, oceanography, and economics is limited. Ocean memory loss impacts not only currents but also crops, markets, migration, and health. We must develop integrative knowledge and co-governance to deal with these cross-cutting threats.

There is no silver bullet. The unknowable can never be fully domesticated, but we can greet it with humility, rigour, and solidarity.

To live in the unknowable is to learn to perceive planning less as blueprints and more as an ongoing practice of learning and adaptation. To prepare not for definite outcomes but for the array of possible worlds. To construct not just systems of technology but systems of trust that can flex and not snap.

The erasure of ocean memory is one of the most significant transformations of our era, not because it warms the planet but because it makes the planet less knowable. Within this vanishing memory, we are faced with environmental problems and epistemological ones: how to behave when we can't anticipate fully, how to project when the past can no longer be used as guidance, and how to hope when the future becomes less clear.

Perhaps there is wisdom in not knowing. A reminder of the fact that the Earth is not something that can be controlled as a machine would be controlled, but something living to be listened to. In the presence of the unknowable, we are not called to commit ourselves to despair but to deeper care, humility, and imagination.

CHAPTER 8

WHO PAYS THE PRICE? FRAGILITY, INSURANCE, AND INEQUALITY

A s oceans become less reliable and climatic extremes become more frequent and unpredictable, the consequences are not shared evenly across the world. The human, economic, and environmental damages are borne unevenly by those who are least responsible for the emergency and incapable of coping with it. In this chapter, we examine the intersection of physical vulnerability, structural inequality, and economic exposure in the case of ocean memory loss and how the disentangling reliability of the ocean lays bare deep structural imbalances at and between social levels.

Climate change has never just been an environmental issue. Climate change is also a reflection of global inequality, and nowhere is it seen more clearly than in the Global South. Nations in Africa, South Asia, Southeast Asia, and certain regions of Latin America are among the most at risk from the capriciousness of the climate yet have contributed the least to accumulated greenhouse gas emissions. When ocean memory dissipates and climates destabilise, these countries are too often inhabited by small-scale farmers, informal workers, and coastal communities on the front lines of an ever-more disordered planet.

The relationship between social vulnerability and ocean memory is multifaceted but powerful. The function of oceans to stabilise temperature, control rainfall, and support sea life has supported entire civilisations. When the function somehow fails, the knock-on consequences spread through food systems, public health, the economy, and the government. For many low-lying and coastal communities where living is inextricably bound up in oceanic rhythms, slight alterations in sea-surface temperature or the patterns of currents can trigger profound instability.

In the population centres of the West African coast, where small-scale fisheries support the livelihoods of millions of people, oceanic temperature and salinity variations, crucial elements of ocean memory, are disrupting fish migration patterns and changing the productivity of vital species. Sardines, anchovies, and mackerel, long-reliable sources of food and income, are moving out of traditional grounds. For communities with minimal other sources of income, this is not only an economic loss but also an existential risk.

In the same way, in South Asia, the diminishing predictability of the Indian Ocean's monsoon cycle, a previously stabilised system supported by ocean memory, has started to threaten agricultural planning. Indian farmers in India, Bangladesh, and Pakistan have come to rely on small windows of opportunity for planting and rainfall reliability for subsistence crops like rice and wheat. Unseasonable rains or extended droughts caused by disrupted ocean-atmosphere dynamics can ruin crops, driving regions towards food insecurity.

These disturbances add to existing vulnerabilities. Most populations hit by these shocks do not have access to strong public infrastructure, credible credit, healthcare, and social safety nets. In these circumstances, small environmental shocks can become humanitarian crises. Since the impacts accumulate, each crop season gone wrong or failed fishery makes each subsequent recovery harder.

Insurance, one of the pillars of risk management in the modern world, is frequently cited as a solution. Theoretically, it has the mechanism for cushioning against risk by dispersing the cost of catastrophic events over time and across more people. However, in practice, the insurance system is not keeping abreast of the risk introduced by ocean memory loss and the general destabilisation of the climate.

In the Global South, access to climate insurance is still scarce. It is either excessively costly, of poor quality, or unavailable to the neediest when it is present. For smallholder farmers or fishers on the economic margin, premium costs can take too large a portion of their income. Index-based insurance contracts, aimed at providing rapid payouts against rainfall or temperature triggers, tend not to detect the finer-grained effects of disrupted climate systems at the local level.

Worst of all, as climatic variation expands, it intensifies the uncertainty of pricing risk for insurers. Reinsurers, those who support the risk assumed by primary insurers, are increasing premiums or withdrawing from coverage in some markets entirely. This "coverage withdrawal" can be seen today in regions such as the Caribbean, where hurricane risk is rising, and in some regions of Africa, where flood and drought cycles do not reliably repeat. Meanwhile, as oceanic conditions become more volatile, actuarial models reliant on experience become less trustworthy, and the business case for ensuring climatic risk deteriorates.

This dynamic is not limited to the developing world. In the United States, Australia, and some regions of Europe, insurers are pulling out of high-risk territory. California has had major insurers exit fire-risk areas, for example, and in Florida, coastal insurance has become prohibitively expensive due to rising seas and more frequent storms. In richer countries, however, insurance pullout leads to policy intervention, state-backed coverage programs, subsidies, or infrastructure spending. In poorer countries, such buffers are non-existent or few.

It speaks to an underlying structural disparity: the unequal ability of states to deal with the new threats posed by climate and ocean system change. States with robust fiscal resources, technological ability, and government institutions have better chances of successfully adjusting to ocean unpredictability. These can afford to invest in resilient infrastructure, enhance their forecasting capacity, and safeguard key industries. Too often weighed down by debt, war, or political instability, low-income nations are reduced to reactive positions, dealing with crises as they occur instead of averting them. These disparities highlight the urgent need for delivering on the promises of SDG 13 (Climate Action) and SDG 10 (Reduced Inequalities), especially through equitable climate financing and adaptation support

A vivid illustration can be found in the experience of Small Island Developing States (SIDS). These countries face double vulnerability: their economic bases of tourism and fisheries are both susceptible to sea warming and acidification, and their existence is threatened by sea-level rise and heightened storm frequency. Weakened ocean memory intensifies these threats by disempowering the capacity to anticipate and prepare for climatic variation. When the sea becomes an unsteady partner, these nations become trapped in a state of perpetual exposure.

It is not geography alone that makes people pay the price. Within nations as well as between them, inequality exists along lines of gender, race, and class. In most societies, women are overconcentrated in agricultural work and informal economies, both of which are enormously vulnerable to the vagaries of the weather. Women also have less access to land rights, credit, and schooling, making it more difficult to bounce back from shocks. In fishing communities, women may be pivotal in seafood processing and selling but are systematically closed out of decision-making circles and management systems for resources. As oceanic conditions become more turbulent, these entrenched gendered disparities can only be expected to increase.

Cities themselves are also impacted. Coastal megacities such as Lagos, Jakarta, Mumbai, and Manila are at the nexus of oceanic and societal vulnerability. Urbanisation has caused informal settlements to develop in exposed flood-prone areas across the city, typically with poor infrastructure and few drains. Coasts are experiencing sea-level rise, storm surges, and tidal flooding, which used to be unusual but now happen more regularly. Millions of city dwellers with little economic reserve or recourse in institutions face these threats. These city systems risk cascading failures when ocean memory decays and extremes intensify.

In response to such inequalities, international climate finance has become essential for redressing climate injustice.

Funds like the Green Climate Fund, the Adaptation Fund, and the Loss and Damage Fund are designed to transfer resources from rich countries to poorer nations to enhance resilience and compensate for impacts. However, they are slow in implementation, underfinanced, and often tangled in red tape. Promises have regularly fallen short of commitments, and bureaucratic barriers and political manoeuvring have regularly held back disbursements.

Furthermore, most of the finance in the climate sector is currently focused on mitigation, curbing emissions over adaptation and loss compensation. Although decarbonisation is essential, it contributes little to confronting the present-day anguish caused by ocean unpredictability. In the words of one Pacific leader, "You can't mitigate a flood after it's destroyed your house." There is a need to harmonise the carbon equation with the human equation: who gets hurt, who gets paid, and who gets to decide.

Another area of potential is the use of community-based risk-sharing and insurance. These rely on local governance arrangements, cooperatives, and customary knowledge systems in designing and offering insurance products. For instance, the parametric insurance structures in Kenya and

Bangladesh have been supplemented with early warning systems, mobile payment systems, and community education to enhance responsiveness and accessibility. Small in scale as they may be, such innovations point to the potential for risk management to be both more equitable and participatory.

However, structural change will require more than marginal innovation. It will require us to rethink how we value resilience, fund adaptation, and quantify vulnerability. Conventional economic systems do not capture the social costs of unpredictability or the returns on preventive investment. Markets systematically discount climate stability until it collapses too late and at exponentially higher costs.

It's at this point the language of justice is crucial. The degradation of ocean memory is a material process and an ethical issue. We're presented with profound questions of responsibility, entitlement, and solidarity. Why should the most exposed individuals be forced to carry the brunt of something they did not cause? What are richer countries' responsibilities to put right, payback, or safeguard? How do we construct global finance, governance, and knowledge systems that account for the power and exposure asymmetries? The responses will not be straightforward. They will necessitate policy changes, imagination, and political will. But the consequences cannot be higher. If we do not recognise and act upon the social dimensions of ocean unpredictability, we risk deepening inequality on a global scale. We risk producing a planet in which resilience is an elite good and vulnerability is a life sentence. Another way exists, though. It starts with recognition of the science and the individuals behind the numbers. Then, it moves forward with accountability from the people in power. It ends in fair action, daring action, guided by shared humanness. In an era of rising tides and short memories, there is one compass alone: justice.

CHAPTER 9

CAN OCEANS LEARN AGAIN? HOPE IN SCIENCE AND TECHNOLOGY

———◆———

T he vision of the ocean as a vast living bank of memories now in the process of forgetting refers to the tragedy and the urgency. It is also an inspiration. For as much as the ocean forgets, we as a species start remembering. Disciplines, borders, and technologies converge on a silent revolution not to restore what has been lost but to learn about, acclimatise to, and live with the sea's changing intelligence. Can the oceans remember again? Perhaps not in the manner of the human brain rewiring itself, but through the new intelligence we apply to their watchfulness and care, some of that memory may be regained.

At the core of this is a constellation of scientific instruments quietly drifting, floating, and orbiting across the planet. Among them, one of the most revolutionary is the Argo float, a fleet of robot sensors dropped in oceans worldwide. There are more than four thousand of them as of the early 2020s. Argo floats dive as deep as two thousand metres, reporting critical details on temperature, salinity, and currents before resurfacing to relay the messages via satellite. These devices have lifespans of several years and cover most of the world's open oceans. They have transformed how we monitor the deep and constantly changing nature of the inside of the ocean.

Before Argo, most of what we knew about the ocean was from the surface or fixed-point measurements from ships and buoys. The ocean's interior was a black box, mostly unknown and under-sampled. Now, with Argo and its extended versions, such as Deep Argo (down to six thousand metres) and Biogeochemical Argo (which also measures oxygen, pH, nitrates, and others), we are starting to assemble an active picture of how heat, salinity, and nutrients are moving through the ocean's vast levels. It is essential to know how the ocean takes in and retains heat, a critical component of its memory, and how these processes are changing under the pressure of climate change.

The Argo is supplemented by an increasing array of ocean buoys in regions of interest. These are joined in their function by the Tropical Atmosphere Ocean (TAO) array in the Pacific, the Prediction and Research Moored Array in the Atlantic (PIRATA), and the Research Moored Array for African–Asian–Australian Monsoon Analysis and Prediction in the Indian Ocean (RAMA). Collectively, they provide real-time sea surface temperature, wind, and humidity. These are vital for predicting phenomena such as El Niño, monsoons, and cyclones. Whenever ocean memory deteriorates, and atmospheric patterns become more unpredictable, these systems become essential resources for regaining some degree of predictability.

Satellite remote sensing, viewing the ocean from space using satellites, provides another layer of understanding. Satellites can also record sea surface height (a proxy for the heat contained in the sea), surface winds, ocean colour (a sign of phytoplankton growth), and sea ice extent. Altimeters on satellites like Jason-3, Sentinel-6, and the future SWOT (Surface Water and Ocean Topography) mission provide scientists with a bird's eye view of ocean currents, ocean eddies, and long-term sea-level rise. Combined with in-situ measurements from Argo and buoys, the information comprises the multi-dimensional mapping of ocean processes in space and time.

All this information would be impossible to manage without the advent of artificial intelligence and machine learning. The sheer quantity of information the ocean observing systems produce daily is much more than what can be handled using conventional statistical techniques. AI can identify patterns, make anomaly forecasts, and even model ocean behaviour in varying climates. These systems are not perfect, but they are getting ever closer. AI-powered models can now be applied to modelling instances of coral bleaching, following the progress of marine heatwaves, and enhancing hurricane predictions after learning from history via decades of observational records.

Significantly, AI is not displacing scientists but complementing their abilities. Machine learning systems assist scientists in processing millions of data points in real-time, freeing scientists to concentrate on interpretation, testing hypotheses, and field verification. This cross-fertilisation of human intuition with computation presents a kind of synthetic intelligence with a distributed memory like the ocean.

A movement of global cooperation supports this widening technological web. The oceans are transboundary; no nation can monitor or manage them in isolation. Acknowledging this fact, institutions like the Intergovernmental Oceanographic Commission (IOC), the Global Ocean Observing System (GOOS), and the World Meteorological Organisation (WMO) have led efforts across multinational partner countries to harmonise data sharing, research priorities, and capacity-building. Initiatives like the United Nations Decade of Ocean Science for Sustainable Development (2021–2030) seek to promote inclusive, cross-disciplinary collaboration between governments, scientists, civil society, and Indigenous knowledge holders. These efforts directly support Sustainable Development Goal 14 on Life Below Water and Goal 17 on global partnerships, recognising that meaningful ocean stewardship depends on shared knowledge, resources, and responsibility.

A case in point is the Blue Planet Initiative, an international collaboration encouraging sustainable use of ocean knowledge for societal gain. It has supported open-access systems, regional centres of learning, and domestic monitoring efforts, particularly in the Global South, where means for observing oceans have long been in short supply. These activities are not merely technical; they are about fairness. In an era where the hazards of ocean forgetting to occur unevenly across the global landscape, knowledge must be shared as well.

On a grassroots level, ocean literacy efforts are on the rise. These efforts seek to take ocean science out of the laboratory and into public awareness. From curriculum in schools to public citizen science projects, communities are gaining an understanding of ocean systems and what is transforming them. Technologies such as "Smartfin" through which surfers can record ocean temperatures using specially designed surfboard fins, and initiatives such as "Seakeepers Society" and "Coastwatch Europe" are engaging everyday individuals in the process of data gathering and ocean stewardship.

Democratising ocean knowledge is essential. With the sea-land interface becoming ever more dynamic and the climatic regime becoming ever more precarious, human societies will have to become ocean-literate not only on the coast but in all its reaches. How the ocean influences Nairobi's rainfall, Punjab's agriculture, or Jakarta's floods is no longer of special interest to anybody; it is everyone's survival issue.

Hope must not be mistaken for complacency. Our capacity to monitor and model the ocean has improved exponentially, yet the warming and acidification do not stop. All it does is delay them long enough for us to get used to them, prepare for them, and, most critically, do something about them. We can monitor the decline of the Gulf Stream using scientific instruments, but we cannot stop it. We can monitor sea-level rise with satellites, but we cannot stop it. That is something we must do ourselves.

The anxiety in the centre of this chapter and maybe of the entire book is the awareness that the more we know about the changing memory of the ocean, the more we realise how susceptible we have become. Technological optimism must go together with moral vision. We can't engineer ourselves out of climate calamity until we fix the underlying issues: dependence on fossil fuels, destruction of habitats, and the political stasis permitting both to go on unchecked.

Amid all such sobering facts, ocean science's work is still providing some sort of radical hope. It is the hope for a new relationship between us and the ocean built not on control and extraction but on knowledge and understanding. The devices we currently deploy in the ocean are not weapons or probes but messengers. They explain narratives never heard, signals from the deep released across centuries of flux. Whether oceans can learn again is also a question about us. Are we capable of listening? Can we remember our interdependence with the oceanic realm we've long taken for granted? Can we develop a new memory across disciplines, languages, and generations with our sensors, programmes, and satellites? The equipment is at hand. The information is accumulating. The sense of urgency is inescapable. Whether we act, heed the sea's vanishing signals and restore the scaffolding of predictability is not only a test of science but of civilisation itself. Ultimately, it is not merely the ocean that forgets. We threaten to forget what living in harmony with the planet means. And it is we who must remember again, learn all over again, and imagine anew.

CHAPTER 10

THE WISDOM OF WATER: LEARNING TO LIVE WITH UNCERTAINTY

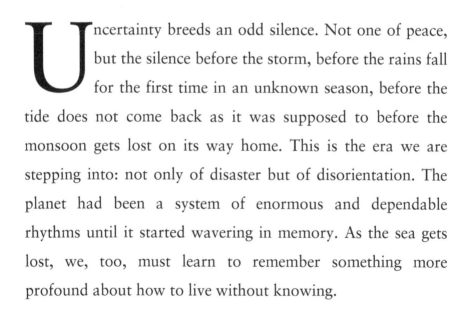

Uncertainty breeds an odd silence. Not one of peace, but the silence before the storm, before the rains fall for the first time in an unknown season, before the tide does not come back as it was supposed to before the monsoon gets lost on its way home. This is the era we are stepping into: not only of disaster but of disorientation. The planet had been a system of enormous and dependable rhythms until it started wavering in memory. As the sea gets lost, we, too, must learn to remember something more profound about how to live without knowing.

Humans have flourished for millennia by seeing patterns. We sowed crops during rains, caught fish during current flows, and settled in cities just out of spring high waters' reach. Predictability has been the support framework of civilisation. The sea, vast and quiet, was a store of much of it. It soaked up the sun's energy, distributed it across the planet, regulated wind and precipitation, and shielded life from extremes. Its record temperature, salinity, and movement were where the seasons themselves resided.

The rules are changing now, and the systems we rely on are becoming unreliable. Oceanic memory is eroding, monsoon patterns are breaking, currents are diminishing, and storms are becoming strangers. This chapter, of all the others, does not presume to explain or analyse. Instead, it inquires what it means to inhabit a world we can no longer reliably forecast, a world in which science, though essential, must be accompanied by humility, wisdom, and an ethic of care.

Water has never been only a chemical compound. It sustains life, shapes landscapes, and governs weather. We depend on it for survival, movement, and balance. The sea is water at its most expansive; vast, powerful, and persistent, yet even it is beginning to change. To live amidst uncertainty is not to abandon science but to redefine it. Science has always progressed through uncertainty. It is a process of honing knowledge of the unknown. But what's new today is the magnitude and immediacy of the unknown. Climate models stretch to encompass ocean dynamics changing faster than projected. Forecasts once good to within several degrees or days now have larger uncertainty buffers. Whole fields are being rewritten in real time. There is no need to be ashamed of it; we must learn to respond.

Adaptation, if based on wisdom, can be truly potent. Think of the people who have always dwelt by water: Pacific Islanders, Arctic Inuit peoples, and Indian fishermen from Kerala to Senegal. To them, unpredictability is nothing new. Oral narratives, seasonally derived knowledge, and some intuitive observational sciences have informed choices before satellites and phone apps. Their knowledge, profoundly empirical and sensitised to the region, presents us with a model, not of resilience against change but of responsive nimbleness.

This does not sentimentalise adversity. These communities are also among the most at risk from ocean change. Rising sea levels, failing fisheries, and altered currents endanger not only their economic bases but also their cultures. Their wisdom, the skill to observe, wait, listen, and adapt, provides something modern systems sometimes do not possess the resilience born of relationships.

Our present way of thinking, particularly in the developed world, has long been predicated on mastery. We have dammed waters, dredged sea bottoms, constructed barriers against waves, and channelled water into the most useful form. Predictability was an artefact of control. But water does not do this. It flows through fingers, seeks the lowest point, goes around things, and wears down foundations in time. It reminds us that lastingness is an illusion.

In the presence of ocean forgetting, we're being called upon gently and urgently to recall our position in the larger web. We're not above the ocean but of it. Its fluctuation is not some distant phenomenon but presents perturbations in farming, energy, migration, psychological well-being, and biodiversity. Ocean memory is unseen by the eye but felt in the texture of everyday life. To do otherwise is to exist in a sort of dangerous arrogance.

So, how do we start living otherwise? The solution, in part, is to adopt another posture, not one of domination but of attention. And it starts with listening to the data, certainly, but also to the people who are already on the front lines, to scientists' warnings of thresholds being reached in their models, and to the elders who narrate stories of broken patterns. It also involves resisting the desire to simplify and cleanse what is complicated and dirty.

Uncertainty begets not paralysis but plural responses. For governments, it creates flexible and inclusive policies that take the slow violence of disrupted climates as seriously as its dramatic catastrophes. It invests in adaptive infrastructure, education, early warning systems, and oceanic research. For communities, it is building diverse knowledge locally, buffering against food and energy insecurity by diversifying sources and reinforcing social networks resilient enough to ride out material and emotional storms.

For people, it requires learning to bear the discomfort and to be in trouble, as author Donna Haraway has described it. It requires reading of shrinking glaciers or a slowed Atlantic current and not being seduced by the siren of denial. It requires discovering rituals of anchoring: walking on the beach, planting trees, showing up at a climate gathering, speaking out at town hall, advocating for policies that acknowledge ecological limits. These are not minor actions. These are the actions through which culture changes.

There is also a spiritual component, if not necessarily religious, to this time. Living in unknown territory demands faith, not faith in the sense of blind belief, but faith in an agential capacity. Faith in the ability to influence what happens in the face of not being in control of everything so that as the planet moves into an era of unpredictability, we are not bereft of the means, knowledge, or action. In this instance, hope is not naivety. It is an application.

What of the sea itself? Despite all its forgetting, it is still vast and generous. It continues to take in carbon, support life, and harbour secrets we have not yet begun to perceive. It is not dying; it is not dead; it is transforming. And maybe in transforming, it is also teaching—not solely of loops of feedback and of thermohaline circulation but of surrender, of cycles and resilience, of what it is to remember another way.

Memories don't always come in sequences. Some reside in sediment strata, coral bands, genes, and ritual cycles of season from hand to hand, from song to song, from tide to tide. The sea does not remember in one way and yet does not forget either. It carries all the shipwrecks, storms, and moments of planetary change in its own flesh.

To live uncertainly means living with this oceanic intelligence in mind. To exchange false certainty for genuine complexity. To prepare not to stay put but to adapt. To mourn, certainly, but also to dream. To write new myths and recount new stories, to have children not for the world we had but for the one becoming now.

This book started in the realm of science with numbers and erosion, models and memory. It closes in something less tangible but real: the potential for becoming more skilled at remembering ourselves in response to a forgetting sea. Not the selves of endless consumption and domination, but those who know to listen, live within limits, and act before it is too late.

Ultimately, the intelligence of water isn't just in its power to sculpt the planet. It's in its unwillingness to be stationary. It flows, it changes, it evolves. It reminds us that not knowing is not the end of information but the start of another. The task is daunting. The risk is high, but the tide is not yet out of reach, let us go forward in hope. The ocean has forgotten some things. It might forget more. We don't have to. Let us remember.

CLOSING THOUGHTS

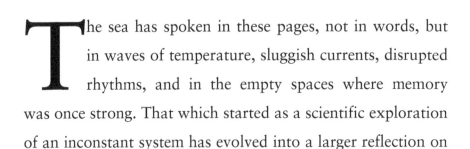

The sea has spoken in these pages, not in words, but in waves of temperature, sluggish currents, disrupted rhythms, and in the empty spaces where memory was once strong. That which started as a scientific exploration of an inconstant system has evolved into a larger reflection on knowledge and where we fit in the natural world.

To know the ocean is losing its memory is not merely to bear witness to the mechanics of global warming. It is to know that we are also impacted in more than material terms. Our economy, food systems, calendar, and conception of time have all been dependent on the sea's rhythms for years. As we watch them disappear, we are being called to become something more than spectators. We are being summoned quietly but unmistakably to remember something primal: our interdependence.

Ocean memory loss doesn't have one solution. It's something with deep roots and spreading echoes. By seeing it and naming it, we can respond through science, policy, education, and cultural change to humility and responsibility.

We can shape a future where uncertainty is not something to fear, but something we are prepared to meet, where our actions are guided not by the short-term, but by the deep time of oceans and ecosystems. We conclude here with an invitation.

Let us remember the sea as a sufferer of change and a collaborator in resilience.

Let us learn not to resist the current but to go with it.

Bindoff, N.L., Cheung, W.W.L., Kairo, J.G., Arístegui, J., Guinder, V.A., Hallberg, R., ... and Williamson, P., 2019. Changing ocean, marine ecosystems, and dependent communities. In: IPCC Special Report on the Ocean and Cryosphere in a Changing Climate. Intergovernmental Panel on Climate Change.

Cheng, L., Abraham, J., Hausfather, Z. and Trenberth, K.E., 2022. Another year of record heat for the oceans. Nature Climate Change, 12(2), pp.92–93.

Intergovernmental Panel on Climate Change (IPCC), 2021. Sixth Assessment Report: Climate Change 2021 – The Physical Science Basis. Cambridge University Press.

Jayne, S.R., Roemmich, D., Zilberman, N., Mensah, V., Gilson, J., Owens, W.B., ... and Wang, H., 2017. The Argo Program: Present and future. Oceanography, 30(2), pp.18–28.

Johnson, G.C. and Lyman, J.M., 2020. Warming trends increasingly dominate global ocean. Nature Climate Change, 10(10), pp.863–869.

Levitus, S., Antonov, J.I., Boyer, T.P., Baranova, O.K., Garcia, H.E., Locarnini, R.A., ... and Zweng, M.M., 2012. World ocean heat content and thermosteric sea level change

(0–2000 m), 1955–2010. Geophysical Research Letters, 39(10).

NOAA, 2020. Global Ocean Heat Content. National Oceanic and Atmospheric Administration. Available at: https://www.ncei.noaa.gov/access/global-ocean-heat-content/

National Research Council, 2010. Ocean Acidification: A National Strategy to Meet the Challenges of a Changing Ocean. Washington, DC: The National Academies Press.

Purkey, S.G. and Johnson, G.C., 2010. Warming of global abyssal and deep Southern Ocean waters between the 1990s and 2000s: Contributions to global heat and sea level rise budgets. Journal of Climate, 23(23), pp.6336–6351.

Rahmstorf, S., Box, J.E., Feulner, G., Mann, M.E., Robinson, A., Rutherford, S. and Schaffernicht, E.J., 2015. Exceptional twentieth-century slowdown in Atlantic Ocean overturning circulation. Nature Climate Change, 5(5), pp.475–480.

Roemmich, D., Church, J., Gilson, J., Monselesan, D., Sutton, P. and Wijffels, S., 2015. Unabated planetary warming and its ocean structure since 2006. Nature Climate Change, 5(3), pp.240–245.

Schmidtko, S., Heywood, K.J., Thompson, A.F. and Aoki, S., 2014. Multidecadal warming of Antarctic waters. Science, 346(6214), pp.1227–1231.

Smeed, D.A., McCarthy, G.D., Cunningham, S.A., Frajka-Williams, E., Rayner, D., Johns, W.E., ... and Duchez, A., 2014. Observed decline of the Atlantic meridional overturning circulation 2004–2012. Geophysical Research Letters, 41(3), pp.610–615.

UNEP, 2021. Making Peace with Nature: A scientific blueprint to tackle the climate, biodiversity and pollution emergencies. United Nations Environment Programme.

United Nations, 2015. Sustainable Development Goals. Goal 13: Take urgent action to combat climate change and its impacts. Available at: https://sdgs.un.org/goals/goal13

United Nations, 2015. Sustainable Development Goals. Goal 14: Conserve and sustainably use the oceans, seas and marine resources.

Available at: https://sdgs.un.org/goals/goal14

United Nations, 2015. Transforming our world: the 2030 Agenda for Sustainable Development. A/RES/70/1. United Nations General Assembly.

Wunsch, C. and Heimbach, P., 2013. Two decades of the Atlantic meridional overturning circulation: anatomy, variations, extremes, prediction, and overcoming its limitations. Journal of Climate, 26(18), pp.7167–7186.

Zanna, L., Khatiwala, S., Gregory, J.M., Ison, J. and Heimbach, P., 2019. Global reconstruction of historical ocean heat storage and transport. Proceedings of the National Academy of Sciences, 116(4), pp.1126–1131.

von Schuckmann, K., Cheng, L., Palmer, M.D., Hansen, J., Tassone, C., Aich, V., ... and Thorne, P.W., 2020. Heat stored in the Earth system: where does the energy go? Earth System Science Data, 12(3), pp.2013–2041.